How Einstein got Relativity Wrong

How Einstein got Relativity Wrong

John Paul Keyes

ISBN: 978-1-952872-98-3

First Printing—2012
Second Printing—2024

Published in the United States by Verge Publishing.
VergePublishing.org

In cooperation with Paul James Keyes

Cover artwork by Paul James Keyes

Internal artwork by Jimmy George

This is a work of non-fiction. Names, places, and incidents are all historical facts.

Preface

According to Einstein's special relativity, *c* is the maximum speed at which all energy, matter, and information in the universe can travel. It is the speed of all massless particles. It is the speed of all electromagnetic radiation, such as light. And, it is also predicted by the current theory to be the speed of gravity or gravitational waves.

Einstein's theory holds that *c* is the speed regardless of the motion of the source or the inertial frame of reference of the observer. In the theory of relativity, *c* interrelates space and time, and appears in the famous equation $E=mc^2$. I am a mathematician and an expert in logic. None of this makes any logical sense—that light can always travel at exactly the same speed and somehow manage to reach all observers at exactly that same speed without taking into account the motion of the observers. How could this concept be accepted by so many and yet be so logically wrong? For years I have ignored it; after all, who really cares whether Einstein's Theory of Relativity is right or wrong? Unless, of course, somebody were to have a more logical theory to offer as an alternative. This book will provide just that—a new theory of reality which adds up both mathematically *and* logically. It will shed light on exactly where Einstein went wrong. Behind us will be the days of light being held constant and time made a variable. Instead, if time is held constant, light is allowed to behave like every other wave. We would no longer need special rules to explain the illogical results of the constancy of light which over the years have led to ridiculous theories of time warps and the possibility of backwards time travel.

Changing Minds

In my view, this mass acceptance of the Theory of Relativity is no different than what Galileo was up against when he tried to change the accepted view that the Earth was at the center of the universe. Galileo was an Italian astronomer and mathematician who played a major role in the Scientific Revolution. Galileo was convinced that the Sun was the center of our Solar system, but needed some sort of evidence. For Galileo, tides provided the physical proof that the Earth was moving. In his theory, tides were caused by the sloshing back and forth of water in the seas as a point on the Earth's surface sped up and slowed down because of the Earth's rotation on its axis and revolution around the Sun. We now know that this theory was wrong. It supposedly would have caused only a single high tide per day. But, Galileo was right about the Earth moving. By 1633 he was ridiculed and even convicted of heresy, because of his idea that the Sun was at the center of the universe and that the Earth revolved around the Sun. This idea was declared contrary to Holy Scripture, as it states "the sun rises and sets and returns to its place," and "the world is firmly established, it cannot be moved." Part of Galileo's sentence included that publication of any of his current or future works was forbidden.

In today's world, there is almost a religious acceptance of relativity. Einstein has been hoisted so far up his pedestal that the thought of someone disagreeing with him has become outrageous. If anyone does, most assume they must be stupid or just don't understand the intricacies of relativity. Not many people think about relativity anymore, and those who do often do not really understand what it is. Many who claim to understand relativity seem to accept it without actually looking at how it got that way or what it really means. Remember, theories are ideas or predictions that typically are difficult if not impossible to actually prove. Theories are generally accepted by the scientific community and by extension the general populace, until someone can disprove them. Then a new theory can take its place.

It is my contention that light cannot travel at a constant speed and still be the same speed for all observers like some sort of magical substance. The "Constancy of the Speed of Light" is a fantasy created to explain our observations. It is quite obvious that light is changing its speed. Light must be systematically changing its speed to match each observer's environment. A ball thrown on Earth travels at the speed of the Earth plus the speed of your throw. In the same way, when light enters our atmosphere it gets dragged along with the motion of the Earth, matching our environment. This very simple concept would explain all our current observations, without resorting to changing clocks and contracting lengths, two results of the math behind relativity.

Unfortunately, I highly doubt that I have any chance of successfully disproving special relativity. Einstein made his theory and the math in such a manner that it actually works in our day-to-day observations. It may likely be very expensive to actually disprove. I do not have access to the type of equipment that would be needed to perform the appropriate types of experiments. The way that science works is this; a theory stands until it is disproved. If you have two competing theories, only the first theory is entertained until it is rejected, then and only then can a second theory be considered. So, what can I hope to accomplish? All I can do is point out the flaws in the thought process that lead to the Theory of Relativity. I hope to get other interested parties to start thinking about what it really means. If enough of the right people look at these concepts, a more logical theory of how light really works could be put in place; the *Theory of Reality*.

It is interesting to note: A cosmic background radiation, discovered in 1964 by American radio astronomers Arno Penzias and Robert Wilson, has apparently shown that Earth may be moving at roughly 386 km/s relative to the radiation. This would suggest that the Milky Way (our galaxy) is hurtling through space at around 600 km/s with respect to the cosmic background radiation. The whole universe is possibly rotating relative to the radiation. With the newest discoveries of neutrinos that may travel faster than the speed of light that have recently come out of the Cern physics labs in Switzerland, it may not be much longer that the theory of special relativity will continue to be accepted. Any particle that can be found to travel faster than the speed of light would specifically refute Einstein's theory.

The idea that time is not absolute; that you change your time and physical size by merely moving in any direction relative to anything else; that nothing can exceed the speed of light; that light always travels in a straight line; that time and space are interrelated; that space is warped by gravitational fields; that you can potentially travel through a worm hole to get to somewhere else; that you can travel backwards in time; all of these thoughts are the ridiculous results of Einstein's Theory of Relativity.

Thought Experiments

Einstein used a number of thought experiments in his description of special relativity to help his readers understand and accept his ideas. I found that those experiments were flawed in their definition and their interpretation. Let me demonstrate a quick failed thought experiment of my own to show you what I mean:

Let's say that you had no concept of vision or light waves. That you were an intelligent bat like creature that used sound waves, traveling roughly 700 mph, to see and measure everything in your world. Your scientists had theorized a constancy of the speed of sound, but you were going to try to disprove it. You think that sound may travel on something, but you are just not sure exactly what it is. So, you decide to search for it by trying to measure a difference in the speed of sound due to your motion in a theoretical *wind*. The current accepted theory holds that sound will always reach an observer at a constant speed of sound (*s*), even if the source of the sound is moving and even if the observer is moving. You are attempting to show that sound actually travels through something. You feel you should be able to measure a *wind* by moving through whatever it is and thus changing the speed of sound, relative to the observer.

So, you build a fancy and expensive speed of sound measuring device. This device breaks a sound wave (that you create) into two pieces sending the waves perpendicular to each other, out a distance and back again. When the two waves come back you combine them and try to detect an interference pattern. An interference pattern is created when the crests and troughs of a wave do not meet at exactly the same place. If you hear any interference this would prove that the sound traveled at a different speed on one leg compared to the other. This would prove that you were moving through a so called *wind*.

Now this fancy piece of equipment (called an interferometer) is pretty delicate. You cannot just pick it up and start flying with it. So, you place this fancy device in a vehicle and start the vehicle moving in a large circle to test different directions. You are surprised to find that none of the measurements that you make show any kind of a difference in speed one direction compared to any other, even though your vehicle was moving. Ta-dah, you believe that you have inadvertently helped confirm that there is no so-called *wind* that affects the speed of sound. You had not considered that sound might be traveling in the air. You apparently had no clue that all of the sound that you created and any sound waves that entered your vehicle from outside would change its speed to match the movement inside your vehicle. Your vehicle was dragging the *wind* with it. If you do not think to place your fancy device outside of your vehicle in the actual *wind*, your measurements and experiments will all be futile.

This is exactly what happened in 1887 with a light wave interferometer. Scientists performed all their tests on Earth. Earth happens to be a closed vehicle in this respect, so they were not able to detect any *wind* with their experiments.

Here is another example of a failed experiment:

Let's start with the same given as above, that you are an intelligent bat like creature. Scientists had accepted a constancy of the speed of sound (*s*) which was fixed at roughly 700 mph. Some bats had suspected that sound may travel faster when they flew higher and higher in the atmosphere. Your task was to show that this could not be possible because sound was always supposed to move at exactly (*s*). So, you build another fancy sound measuring device. This device, of course, is too heavy for you to carry and besides you need it to bounce sound off of an opposite wall at a known distance and back again in order to measure the speed. So you place your fancy device in a really tall building that has walls that are equidistant apart all the way up. You measure the speed at the bottom floors, then you measure at the top floors of the building. Both locations measure the speed to be exactly the same. Ta-dah, you believe that you have confirmed that altitude does not affect the speed of sound. Oddly, that just happens to be true. What you did not consider was that the speed of sound might actually be affected by the temperature, not the altitude. In an outside environment, the temperature typically goes down as you move up and you would have found a difference in the speed at higher (colder) elevations. But, you did your fancy experiments in a temperature controlled building and found nothing. When you don't know exactly what you are looking for, you may find unexpected or irrelevant results.

If you take the results from these last two bogus experiments, you will be using the following incorrect results: First, that the speed of sound is constant regardless of the motion of the observer. Second, that the speed of sound is constant even at different altitudes. Using these incorrect or misinterpreted results your mathematicians are ultimately going to end up with formulas showing that both time and distance must change with motion and altitude. There is just no getting around it. It will also be shown that if you travel at a high rate of speed in your vehicle, that your time will have to be slower—therefore you will age slower than your other bat friends.

As ridiculous as this seems, this is exactly what is predicted to happen with the Theory of Relativity. As any object travels close to the speed of light, its clock will theoretically run slower. Twin brothers, one in a spaceship traveling far away and the other left on Earth, would supposedly age differently. But, who really is moving—the one in the spaceship or the one on Earth?

All motion can only be measured within a frame of reference. When two or more frames come into play at once, there can be multiple interpretations of motion relative to each frame. No one can say what is actually moving, as there is supposedly no *center* of the universe. Two satellites hurtling side by side around the Earth at the same rate would view each other as stationary. They can only interpret their motion by watching the Earth rotate below. The solar system is moving as well, as is the galaxy as a whole, making all motion of the two satellites compared to the Earth just as relative as the satellites to each other or any other frame of reference.

In the case of the twin brothers, one moving "away" quickly, each could consider the other brother to be the one moving relative to their own environment. Because we use light (running at a finite speed) for our measurements, each of the brothers would observe the clock of the other to be running slower. This is simply an observational mirage, as light takes longer and longer to reach each observer as the distance between them rapidly increases. In the same way, the observed clocks would appear to run faster as the two approached each other. Light would arrive sooner and sooner, making it appear that the clocks were ticking faster.

Time has not actually slowed down or sped up—that false interpretation is just the product of the light of the observed clock stretching out before it reaches the eye of the observer. As the brothers come back together, the light is propagated on top of itself so that from either brother's frame of reference, the other brother's light reaches them more rapidly than it was actually generated, giving the illusion of time speeding up. Once we hold time as a constant rather than light, much more logical solutions present themselves.

Back in the intelligent bat's world, nothing could travel faster than the speed of sound (s). The basic problem that you have is that you are relying on sound (with its inherent limitations) to do all of your measurements. You are also ignoring the fact that the speed of sound may actually be controlled by something—just because you don't know what it is.

If something did move faster than sound, you would have no way of observing or measuring it, or would you? Let's envision what would happen if an object did move right past you, going, let's say, 900 mph. What do you think that you would witness, what would it *sound* like to you?

Here is what you might observe: As the object was approaching you, you would not be able to detect it. It would be traveling faster than your sound waves or its own sound waves could reach you. It would arrive right in front of you, even before its own sound got to you. Just as it arrived right in front of you, you would suddenly hear it. It would supposedly appear out of nowhere right in front of you. And, here is the odd part—it would appear to explode into two pieces; an object and an anti-object apparently moving in opposite directions.

The real object would actually scream off going in the original direction not quite at the full speed of sound. You cannot hear objects moving faster than that, as the sound takes this long to reach you. It would appear to be moving at close to your top observational speed (s), even though it was actually moving faster. You would theorize that the object's time was running slower than yours. If there was a clock on the object, the clock would appear to slow down to almost nothing.

The anti-object would spontaneously appear and move backwards in the direction that the actual object came from. It would look like a negative version of the original. As the approaching sound waves started catching up to your location, the anti-object would appear to move away from you at a high rate of speed. Faster than it should be able to move. As you can see, using something with a finite speed (sound waves) to do your measurements leaves you with limited observational capability. Weird math that fixes the speed of sound but changes time and distances can try to make sense of your observations, but it does not explain what is really happening.

Does any of this sound familiar? Antiparticles can be generated in a particle accelerator by radioactive decay, or by cosmic-ray collisions. They were first discovered by Carl Anderson in 1932 when he studied the trails of particles left in a cloud chamber. At first they were supposedly mistaken for an electron going in the opposite direction, but later they were determined to be positrons. Positrons are defined as anti-electrons that have a positive charge causing them to rotate in the opposite direction from electrons. Particle-antiparticle pairs will annihilate each other, if they come in contact. They are created simultaneously in particle accelerators, bursting away from each other traveling in opposite directions.

Background

Now back to our task of exploring relativity. Let's start by getting some background into what information Einstein and his predecessors had to work with.

The Speed of Light was first determined to be finite by Ole Romer, a Danish astronomer in 1676. He used measurements from the eclipses of Jupiter's moon Io, to make this discovery. Actually Giovanni Cassini, Romer's teacher, had observed the same eclipse 10 years earlier and had seen the same discrepancies in his measurements. At first, he attributed the discrepancies to light having a finite speed. They both noticed that the times between eclipses got shorter as Earth approached Jupiter, and longer as Earth moved farther away.

Oddly, Cassini seems to have abandoned this reasoning, which Romer adopted. Romer never actually calculated the speed of light, however many others did using his data. Romer's view that light had a finite speed was never fully accepted by mainstream astronomers until 1727 by James Bradley.

Bradley was attempting to measure a stellar parallax when he noticed that some of his measurements appeared way off.

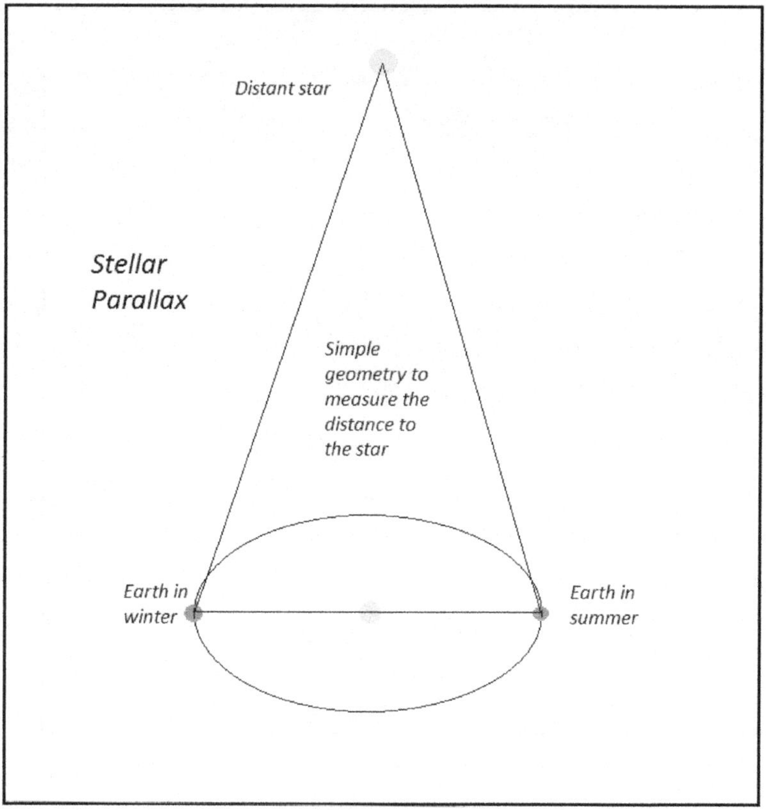

Stellar parallax can be used to determine the distance to a star from the Earth. The parallax is created by different orbital positions of Earth as it rotates around the Sun. By measuring the angles, one can determine the distance to the object. Bradley was attempting to calculate a stellar parallax when he stumbled onto or discovered a stellar aberration.

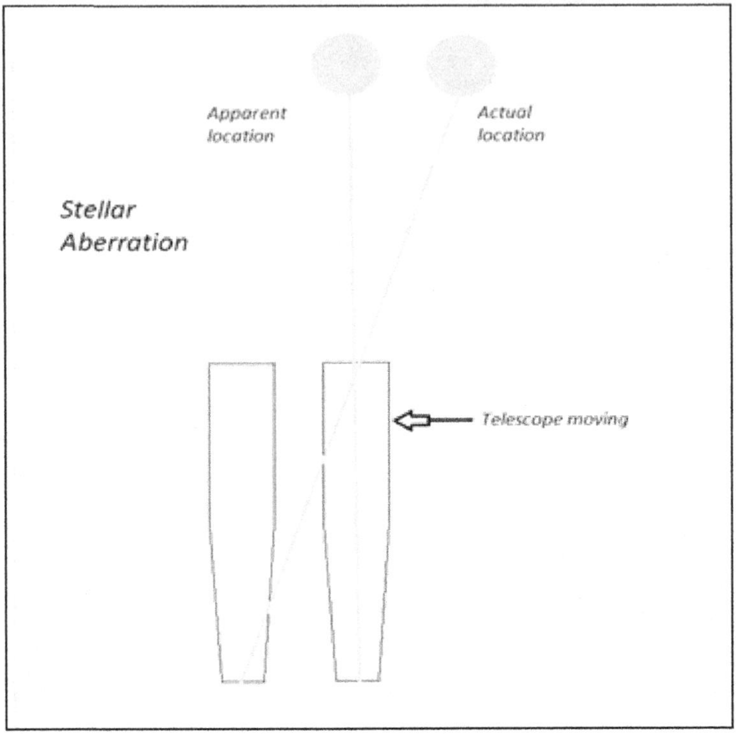

A stellar aberration occurs when the Earth moves through space horizontally or perpendicular to the observed star. Bradley noticed that the star appeared in the wrong place when Earth was moving its fastest, left or right, relative to the star. With a bit of thought, he finally figured out what it meant when he compared the results to his sailing experiences. When a sail boat moves sideways through the wind, the apparent wind is shown by an indicator usually located at the top of the mast. That indicator does not point to the actual wind—it is shifted forward relative to how fast the boat and the wind is moving. This is an aberration.

This finally proved that light was traveling at a finite speed. The speed of light in a vacuum, which is usually denoted by *c*, has been calculated at approximately 299,792 kilometers per second (km/s). The speed of light through other transparent materials, such as glass or air, is less than *c*.

The Doppler Effect or Doppler shift was named after an Austrian physicist Christian Doppler. In 1842, Doppler described a change in frequency of a wave for an observer that was moving relative to the source of the wave, or for the source moving relative to the observer. It is commonly heard when a vehicle making a sound approaches, passes, and recedes from an observer. The received frequency is higher (compared to the original frequency) during the approach. It is the same as the vehicle passes by, and it is lower as the vehicle goes away from the observer. Light does the exact same thing. The Doppler Effect with light is often referred to as either blue shifted when approaching or red shifted when receding relative to a light source.

Electromagnetic Waves

Michael Faraday, an English chemist and physicist, was an experimentalist in the first half of the 19th century. Faraday researched magnetic fields and electric currents to discover that magnetism could affect rays of light in a medium and that there was an underlying relationship between the two phenomena. His inventions of electromagnetic rotary devices formed the foundation of electric motor technology.

James Maxwell took the research of Faraday, and others, to formally postulate electromagnetic waves. Electromagnetic waves can be imagined as a self-propagating transverse oscillating wave of both electric and magnetic fields. If the electric field is oscillating, let's say up and down, then the magnetic field will be oscillating left and right. Maxwell showed that a changing magnetic field induced an electric field, and that the changing electric field induced a magnetic field. Therefore, electromagnetic waves were possibly self-sustaining as they traveled through empty space. Maxwell's equations predicted the speed of electromagnetic waves, as a constant value (c), which happened to coincide with the speed of light. Because the speed of EM waves coincided with the measured speed of light, Maxwell concluded that light itself is an EM wave at certain wavelengths. Using the data available at the time, it was calculated at 310,740 km/s. A wave is characterized by a velocity, wavelength, and frequency. Whenever you measure the velocity of an electromagnetic wave, it always appears to propagate at the same speed, the speed of light (*c*).

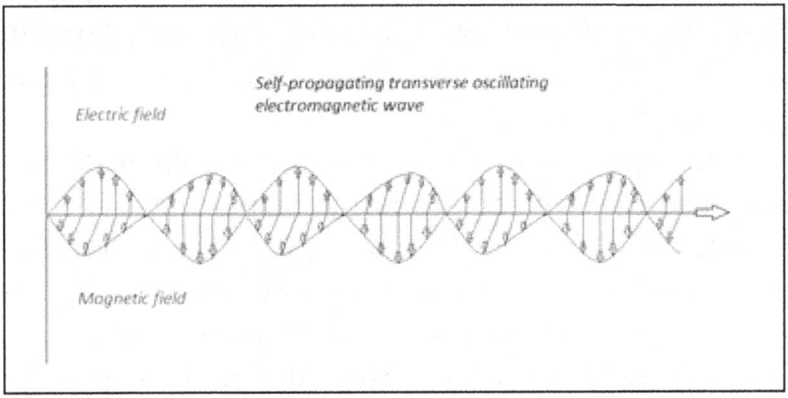

It is said that Einstein kept pictures of Isaac Newton, Michael Faraday, and James Maxwell on his study wall.

Luminiferous Aether

The Fizeau experiment: In 1851 Hippolyte Fizeau did a number of experiments on the speed of light. In one of his experiments, it was found that the speed of light was affected by traveling through a moving liquid. Michelson and Morley successfully repeated that experiment in 1886.

Earth rotates at around 1000 mph or about .5 km/s. It is also moving rapidly through space at a speed of around 67,000 mph or about 30 km/s as it rotates around the Sun. Our Sun itself is traveling around 492,000 mph or 220 km/s as it rotates around the Milky Way (our galaxy). Since the Earth is obviously in motion, it was expected that we should be able to detect a difference in light speed in certain directions. A lot of experiments have been done to try to find this theoretical difference in light speed. All have failed. So, how is it that light will still hit an Earth bound observer at the same speed, even though the Earth is rapidly moving through space?

For many years scientists looked for a luminiferous aether. This was the theoretical medium through which light propagates in empty space. Light has wave properties, therefore it was thought that it must be traveling through something. If an aether existed, it was theorized, we should be able to measure our movement through the aether. We should be able to find an aether wind as Earth travels around the Sun through this medium. Michelson and Morley did a famous experiment in 1887, where they created an interferometer to detect the aether wind. For years they refined the equipment, but were never able to

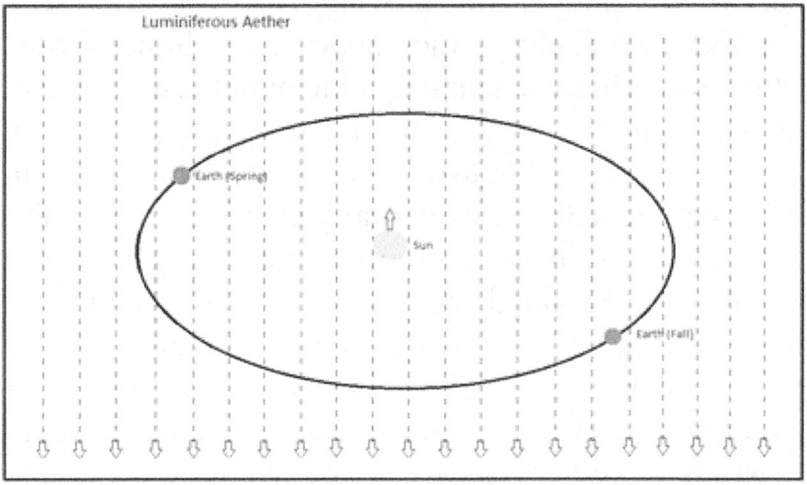

detect any aether.

Galileo defined a principle of relativity which states that all mechanical laws of physics are exactly the same for every inertial observer. In other words, when you are in any type of closed vehicle or system, you would have no way of knowing if your vehicle is moving or not. Everything else outside your environment might be moving, but as far as you are concerned your environment is stationary. The Earth itself, in this context, is a closed system and an inertial frame of reference. Unfortunately, Galileo's Spacetime was incompatible with the notion of an aether. If an aether existed, it might be possible to determine absolute motion relative to the aether and therefore the principle of relativity would be disproved. An astute observer would be able to tell that his environment was moving.

In the 19th century, the theory of a luminiferous aether was widely discussed. One hypothesis was that the aether was somehow entrained or dragged within moving matter. Augustin-Jean Fresnel proposed in 1818 that the aether was partially entrained by matter. George Stokes in 1845 proposed that the aether was completely entrained within or in the vicinity of matter. Stokes' complete aether dragging model could explain the negative outcome of all aether drift experiments. However, the theory was considered to be wrong for other reasons. Experiments showed that light was not affected by large objects rotating or placed near their measuring equipment. Stellar aberration (what James Bradley had discovered in 1727) was also not considered possible if the aether was dragged with the Earth. Light would fall straight down a telescope rather than at an angle as Earth carried the telescope through space.

Theodor des Coudres and Wilhem Wien proposed another version of Stokes model in 1900. They assumed that aether dragging was proportional to the gravitational mass. That is, the aether was completely dragged by the Earth and only partially dragged by smaller objects on Earth. To explain aberration, Max Planck in 1899 argued that the aether might not be incompressible, but condensed by gravitation near Earth. This theory was refuted by the Michelson-Gale-Pearson experiment in 1925. They used a large set of mirrors that was sensitive to the rotation of Earth to bounce light around. If the aether was completely dragged by the Earth's gravitational field, they expected a negative result, but the result was positive.

Local Time and Contraction

In 1892 and 1895 Hendrik Lorentz created a new theory based on a completely motionless aether. It explained the failure of the negative aether drift experiments by introducing an auxiliary variable called "local time" for connecting systems at rest and in motion in the aether. In addition, he introduced a "length contraction." By 1899 and 1904 he expanded his theory by introducing the Lorentz transformation. Lorentz based his theory on the aether theory of Fresnel where the aether was partially entrained by matter. A big challenge for this theory was the Michelson-Morley experiment done in 1887, which showed no aether movement. It was as if the aether was fully dragged by Earth, and yet this would prevent the effect of aberration.

In an attempt to explain the immobile aether results, George FitzGerald in 1889 suggested that all electrostatic fields and molecular forces were affected by their movement. Everything would contract in the line of the motion. Lorentz independently hypothesized the same contraction in 1892. He included that an observer co-moving with the Earth would not notice this contraction, because they would also contract at the same rate. Any measuring device would shrink as it was turned toward the line of movement and expand when turned perpendicular. How convenient.

An important part of Lorentz theorem was the "local time," which was different for observers in different frames of reference. By adjusting the "local time" Lorentz was able to help explain the aberration of light, the Doppler Effect, and the Fizeau experiment. He considered the time transformation only as a "heuristic working hypothesis and a mathematical stipulation to simplify the calculations from a resting and a 'fictitious' moving system."

Contrary to Lorentz, Henri Poincare saw more than a mathematical trick in the definition of local time. He called it Lorentz's "most ingenious idea." Poincare then goes off on a tangent. In 1898 he writes a paper that is called "The Measure of Time." He explained that astronomers, in determining the speed of light, simply *assumed* that light had a constant speed and that it was the same in all directions. He felt that without this postulate it would be impossible to infer the speed of light from astronomical observations. In his paper he uses a lot of double talk, hand waving, and thought experiments to demonstrate that time is difficult to measure, especially if you are attempting to compare it in two different places. Here is a small piece of what Poincare wrote:

"We do not have a direct intuition for simultaneity, just as little as for the equality of two periods. If we believe to have this intuition, it is an illusion. We helped ourselves with certain rules, which we usually use without giving us account over it [...] We choose these rules therefore, not because they are true, but because they are the most convenient, and we could summarize them while saying: The simultaneity of two events, or the order of their succession, the equality of two durations, are to be so defined that the enunciation of the natural laws may be as simple as possible. In other words, all these rules, all these definitions are only the fruit of an unconscious opportunism."

He goes on and on and even on some more about it, until he convinces himself that there is no absolute time and that it is reasonable to have two different times in two different frames of reference. And that those two clocks do not measure the same time. Why is he doing this? By altering time he can explain all of the failed aether experiments. If you cannot get your math to work with your current theory, then try changing the variables so that things don't have to add up.

Poincare interpreted local time as the result of a synchronization procedure based on light signals. In other words, two observers would synchronize their clocks by optical signals. Since they believed that they were at rest within the aether, they could consider only the transmission time of the signals in determining whether their clocks were synchronous.

Relativity

In 1895 Poincare argued that experiments like that of Michelson-Morley suggest it is impossible to detect the motion of matter or the relative motion in relation to the aether. Although most physicists had other views, Poincare in 1900 stood to his opinion and used the expressions "principle of relative motion" and "relativity of space." He criticized Lorentz by saying that it would be better to create a more fundamental theory, which explains the absence of any aether drift, than to make one hypothesis after the other. In 1902 he used for the first time the expression "principle of relativity."

Poincare then introduces the modern "principle of relativity" by introducing more relativistic velocity transformations into Lorentz transformations by 1905. This was an important step in the formulation of the theory of "special relativity." The principle of relativity stated that we have no means, and can have none, of determining whether or not we are being carried along in a motion. All the laws of physics must be the same for a stationary observer as for one moving along in a different frame of reference.

In 1908 Walter Ritz, a Swiss physicist, produced a lengthy criticism of the Maxwell-Lorentz electromagnetic theory. He contended that the theory's connection with the luminescent aether made it "essentially inappropriate to express the comprehensive laws for the propagation of electrodynamic actions." Ritz pointed out seven problems with Maxwell-Lorentz electromagnetic field equations. Ritz had a competing theory for the special theory of relativity called the "emitter theory" or "ballistic theory of light." Emission theories obey the "principle of relativity" by having no preferred frame for light transmission. Light, they say, is emitted at speed (*c*) relative to its source. This would explain the results of the Michelson-Morley experiments, because they created the light that was being tested. This theory is considered by most scientists as discredited, because of the lack of extreme optical behavior of binary star systems. The light from the approaching star might be expected to travel faster than light from its receding companion, and overtake it. Numerous terrestrial experiments have also been performed that demonstrated that light speed was independent of the speed of the source.

Finally, Einstein, in 1905, published the "special theory of relativity" where he abandoned completely the idea of an aether. He fully accepts the constancy of the speed of light. Most scientists have now accepted Einstein's FitzGerald-Lorentz contraction, or simply called length contraction. According to this physical law all objects physically contract along the line of motion. This is used to explain how light would transit slower on one arm of an interferometer, but ends up traveling a shorter distance that exactly cancels out the drift. Einstein's special relativity is generally considered the "solution" to the Michelson-Morley null result. As late as 1920, Einstein himself still spoke of a different concept of aether that was not a "ponderable medium" but something of significance nonetheless.

What no one had figured out was that all of the aether wind experiments had been done on Earth, within the gravitational field of Earth. If an aether wind does exist, and if mass or gravity has anything to do with it, there should be no way that we can detect it here on Earth, inside our vehicle (Earth). According to Maxwell's equations on electromagnetic waves, light can propagate itself through empty space. It apparently does not need a medium to keep it going. Light is after all made up of a changing electric field that creates a changing magnetic field that creates a new electric field which in turn creates a new magnetic field and so on.

But, logic dictates that there must be some sort of a luminiferous aether. Just the fact that light propagates on Earth at a constant speed suggests that something is causing this. We just have not figured it out yet. If we have any chance at all to find an aether wind, we would have to put our measuring device outside of our vehicle (Earth). We would basically have to travel deep into empty space, away from large gravitational fields, to find it. We also should know a little bit about what might make up the aether, so that we would know where or how to look for it.

As a recap: Michelson and Morley performed repeated experiments to find an aether wind, but did so on Earth, producing null results. Then along came George FitzGerald. In 1889 he published a short paper "The Ether and the Earth's Atmosphere." In this paper he conjectured that if all moving objects were foreshortened in the direction of their motion, it would supposedly account for the curious null-results of the Michelson-Morley experiment. The Dutch physicist Hendrik Lorentz hit on a similar idea in 1892 and developed it more fully into Lorentz transformations. Two observers must vary their measurements of space and time in order to understand each other's frame of reference. This was used to explain the mystery behind why observers moving at different velocities may measure different distances, elapsed times, and even different ordering of events. To be fair, Michelson went to his grave convinced that light speed was inconstant in different directions, and also convinced of the existence of an aether.

Let's look at what this "length contraction" really means. In the following illustration, two observers A and B are each moving in different directions from an object. Each will see that object contract in the direction of their movement. Observer A considers that the object is contracted from top to bottom, whereas observer B considers the same object is contracted from left to right. What are we supposed to make of this?

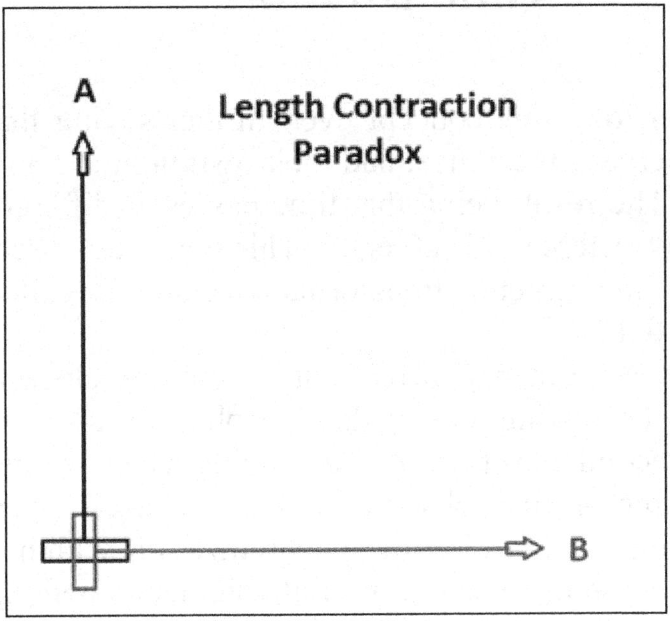

Time Dilation

Einstein took this concept even further saying that each observer, in essence, had their own unique set of clocks. The result being that time passes at different rates for different observers. This was the direct result of the Lorentz transformations and is called time dilation.

In the special theory of relativity, a moving clock is found to be ticking slower than an observer's clock. But, a second observer at the moving clock would also see the original observer's clock ticking slower. There comes a question of synchronization within a reference system. The question of whether something happening in one system is in fact happening simultaneously with something elsewhere. Establishing simultaneity of events separated in space necessarily requires transmission of information between locations. This indicates that the speed of light will enter the determination of simultaneity.

Einstein described a thought experiment concerning simultaneity that Poincare had originally described. He pictured two frames of reference which included a long train moving past a stationary platform. Just as the observer M' in the train passes by an observer M on the platform, lighting strikes at either ends of the platform A and B, at equal distances from observer

M. If the observer M on the platform sees both lighting strikes at the same time, he will conclude that the lightning strikes were simultaneous. The observer on the train M’, however, is moving. The train observer will detect one strike B prior to strike A, as the observer moves relative to the platform where the strikes originated. This, as Einstein concludes, means that events are only simultaneous for observers within their own frames of reference and will not be simultaneous for other observers in other frames of reference.

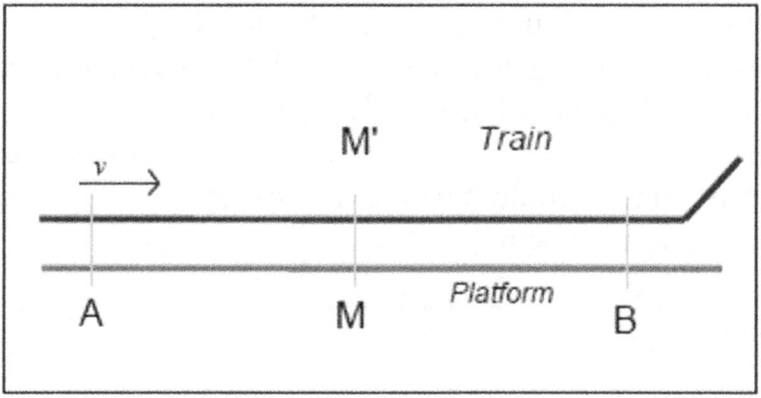

But what if you have an intelligent observer in the train? That intelligent observer would recognize that all observations are recording past events, because the speed of light is finite. If that observer could determine that light from each of the strikes was being propagated at a certain speed (c) relative to a specific inertial frame, such as the platform, he could do some quick calculations. With just a little bit of extrapolation, the intelligent train observer can take the time that each lightning strike had been recorded, and using the speed that the train was moving relative to the platform, he could determine how far he had traveled since each strike had originally occurred. He could then potentially determine that both strikes had originally happened when he was opposite observer M on the train platform and therefore both had originally happened at exactly the same time. An intelligent observer would figure out that light traveling within the frame of reference of the platform was not traveling at exactly (c) relative to the moving train. Of course, that intelligent observer would also note that any light that entered the frame of reference of the train would change its speed to match the train's environment.

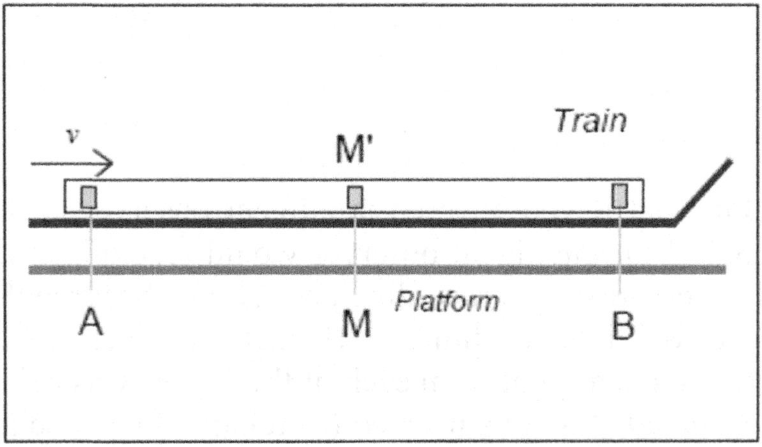

Let's take Einstein's thought experiment a little further and place windows at both ends of the train, just opposite A and B. Any light that entered the train through those windows would change its speed to match the train's movement. This means that the light that observer M' detects that traveled within the frame of reference of the train would actually be simultaneous, even though observer M' has moved by the time the light reaches that point. Knowing what frame of reference that light is traveling within, is key to understanding reality.

Let's say that an observer was located on Jupiter. I have no doubt that if you were to measure the speed of light as it was hitting that observer, from any light source, that it would be traveling at some speed, let's say the speed of light (*c*). And, at the same time, the light that hit an observer on Earth, from any source, would also be exactly *c*. If both observers were to measure the light coming from the same source, let's say some star, both would detect the light coming from that source at exactly the same speed, namely *c*. And, yet both observers and even the light source are all moving relative to each other. How can light, from the same moving source, hit both observers at the same speed? You have two choices: You can either accept Einstein's special relativity which calls for any movement to magically change time and distances so that light always has the same speed. Or, we can look for a more logical explanation behind our observations. I choose the latter.

Binary Star

Let's picture in our mind a binary star. We already know that the light that reaches us from both parts of the star must be traveling at the same speed, otherwise we should detect some odd optical effects. However, it is also accepted that light that emerged from each of those stars was traveling at the speed of light as it left them. When the source of the light has a speed of say u (because of its rotation), logic would dictate that the speed of the light emitted in the direction of travel would have to be $c + u$. And, the speed of the light from the other (receding star) would have to be $c - u$. Yet, by the time the light from each of those sources reaches us, it is traveling at a constant speed of c. This is one of the observations that have been declared a proof for the constancy of the speed of light. Everywhere you look, light seems to be traveling at the same speed, even when the observer is also moving (with Earth).

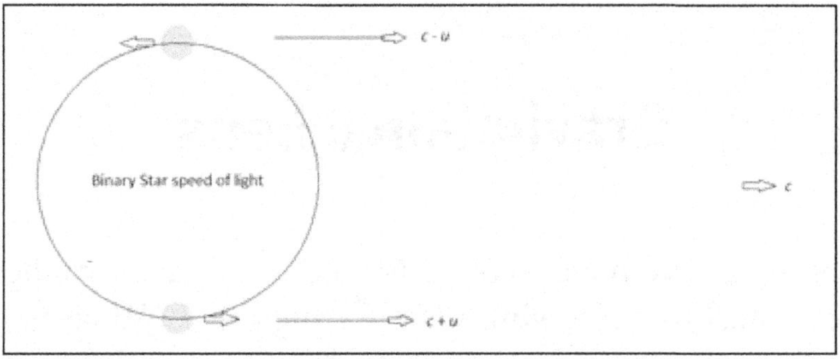

Think about that binary star for a moment. An observer at either one of those stars sees the light leave at exactly *c*. What happens? How does the light change its speed to match the light coming from the other star? Logic dictates that there must be some sort of an aether. Something must cause the light from both of those sources to somehow change their speed to match each other. That 'something' must be the aether that we have been so desperately looking for. If you are looking for evidence of an aether you don't have to look any further than this to realize that it must be there.

Gravitational Lens

Now, all we have to do is to find this illusive aether. Let's start by reviewing what we do know. What is in empty space? If you were to take a clear box into empty space and seal it up, what would be left inside the box? First we can see through the box, so we know that light waves are traveling through it. But there is still something else left in the box that you cannot get rid of: Gravity. Gravity goes through everything. Its force is felt inside the box and through the box.

What affects light? This is easy. We already know that gravity affects light. We know that the stronger a gravitational field is, the more that it slows light down. In extreme cases, we know that beyond an event horizon of a black hole the very strong gravity will prevent light from escaping.

A gravitational lens refers to a distribution of matter between a distant source and an observer, that is capable of bending (lensing) the light from the source.

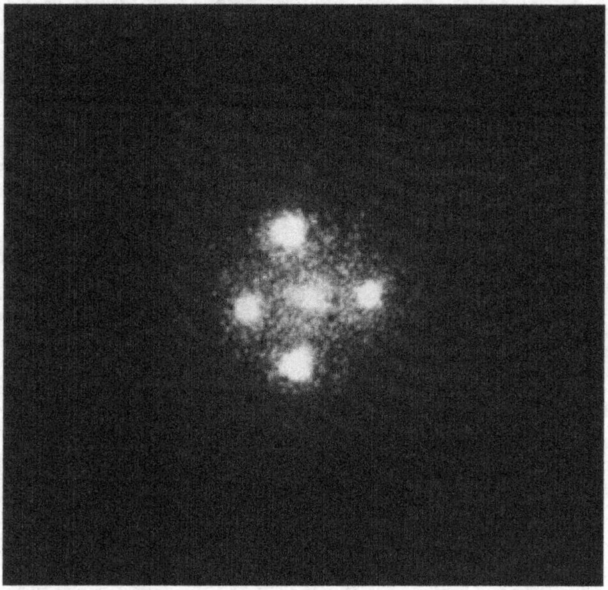

Look at this picture of a single quasar appearing to be in four places at once around a single foreground galaxy due to strong gravitational lensing. This is referred to as the Einstein Cross.

Look at how this massive cluster of galaxies causes this large lensing effect. Eerily distorted background galaxies surround the cluster. The gravity of the cluster's trillion stars acts as a 2 million-light-year-wide lens in space.

This image shows a galaxy cluster with an enormous mass, powerful enough to visibly bend the path of light, like a magnifying glass. These lensing clusters are useful tools for studying very distant objects. The lensing behavior amplifies the light from the faraway galaxies in the background. Hint: The further away a galaxy is the redder it will appear.

A gravitational lens is already known to bend light from a source coming from behind it. Einstein's General Theory of Relativity predicted this lensing effect. But, it supposes that the gravity has warped or curved space causing this effect. It should be obvious that strong gravitational fields are actually slowing down the light. The stronger the gravity, the slower the light will be moving. When any wave is slowed down, in this manner, the wave will bend toward the slowing source. Light waves are no different. They are obviously bending as they pass by any strong gravitational field. This slowing down and bending has no limit. Light that gets too close to an extremely strong gravitational field, like a black hole, would therefore bend itself right around in a circle, such that it would not escape. The cluster of galaxies shows that the slowing down effect stretches out two million light years. This means that light is traveling faster the further it gets from a strong gravitational source. How must faster? No one really knows.

The Speed of Light

As far as I have been able to determine, no one has actually measured the speed of any non-terrestrial light source. Each of the aether tests and light speed tests were performed using a source of light that was generated on Earth. The light that we generate obviously travels with the speed of the Earth as it moves through space. This must have played a part in why so many people believed the Emitter theory for so long. We revolve around the Sun at pretty much the same distance year round. We have no clue what the speed of light would be from the Sun, if we were moving toward or away from it. It's never been measured.

Scientists are using electromagnetic waves, including light, to do all of their astronomical measurements. Not knowing how fast light moves amongst the stars was a conundrum. Astronomers, mathematicians, and physicists made the conscious decision to assume that light had a constant speed, and that this speed was the same in all directions. Without this assumption, it was thought that it would be impossible to infer the speed of light from astronomical observations. When Ole Romer observed the moons of Jupiter, he had to make an assumption about the speed of light in order to make any sense of his measurements.

So here is where we are: Because we do not know what may cause light to travel at any particular speed, we decide to wave our hands and pronounce that light has to be constant from all directions everywhere. Worse yet, it was then decided that even if an observer is moving, the speed of light is still going to be the same. This assumption, which they call a stipulation or a postulate, is exactly what has caused all of the ridiculous problems such as time dilation, and width contraction. It forms the basis for special relativity. None of the scientists, especially Einstein, were actually describing physical reality, only an ideal reality that is 'imaginarily' perfect. They were all trying to describe how we can effectively measure things in the known universe, using the visual tools that we have at our disposal. It is like a teacher giving everyone a ruler and asking them to measure an object, each student must make the assumption that their ruler is the same length as the other rulers. Otherwise, their measurements may not have any meaning to anyone else. But, declaring the constancy of light will cause your other tools, namely your clock and your measuring stick to have to change with your movement. If light is considered constant then time and length cannot be fixed. It's one or the other, you cannot have both.

The physicists and mathematicians of the 19th and 20th centuries decided to accept the "Constancy of the

Speed of Light." What was their evidence? The first piece that is usually talked about is binary stars. The fact that no abnormal optical properties are observed supposedly proves that light speed is not affected by motion from the light source. In actuality, binary stars demonstrate that light cannot be constant. Light must change its speed in transit to match each other, otherwise we would no doubt observe some strange optical behavior of a binary star. The fact that we don't see strange behavior clearly shows that light must be changing its speed.

The next piece of information that one might think of is the fact that we cannot measure any aether drift here on Earth. This is supposed to prove that light is constant regardless of the motion of the observer. It is my belief that this again actually demonstrates that light cannot be constant. Earth is not the center of the universe. Light must change its speed in transit to reach Earth at a fixed speed. There just is no way for light to leave a moving source at some fixed speed, travel at that same speed and reach another moving target at that same fixed speed again without changing its speed in transit. Let's get real here for a moment. Light is not magic, it is simply an electromagnetic wave. The very fact that light is observed to reach Earth at a fixed speed should be—to any logical, intelligent person—certain proof that light must have changed its speed to match the movement of Earth at some point during its journey through the cosmos.

Once you accept the absurd notion of the "Constancy of the Speed of Light," you are doomed to have everything else wrong. This is exactly what Einstein and all the other mathematicians before him have done. In order to make any observational calculations work, you will have to assign a different clock to every system that moves relative to your location. You have to make every object contract in the direction of their movement. And, you have to put limits on everything because your math will not work beyond this theoretical maximum speed of light, just like the intelligent bat's math could not work beyond the theorized constant speed of sound.

Reality

Let's look at what we are attempting to explain, for a Theory of Reality:

1) Light generated on Earth propagates at the same speed in all directions, even though Earth is moving through space.
2) Light generated by the Sun propagates the same in all directions, even though the Sun is moving through space.
3) Light generated by a binary star reaches Earth pretty much at the same time, from both legs of the binary source, without causing any strange optical behavior.
4) When Earth moves perpendicular to a light source we can detect a stellar aberration.
5) Light bends near a strong gravitational source.
6) If you move a light source on Earth, this does not affect the speed of the light for a stationary observer.

We already know that gravity is affecting the speed of light. It is my postulate that gravitational fields are responsible for all of the measurements of light that we observe. I am proposing that a gravitational aether is what everyone was looking for. For many years,

Einstein had looked at this very prospect, but nothing ever came of it. It was assumed that gravity would have to drag the light with it, which would prevent stellar aberration. However, it does not act like the type of aether that they were expecting. It does not constrain or entrain light. Light does not ride on the aether. The aether (gravity) instead acts something like a governor for the speed of light in empty space. The emitter theory (where light travels at the speed of the source) is partially correct. Light that we generate here on Earth demonstrates that. The Michelson-Morley experiments demonstrated that light generated by a terrestrial source travels with the movement of the Earth, preventing any aether detection. Earth was not dragging the aether, it was producing it. Other terrestrial experiments have shown that light is not affected by a source moving relative to the Earth. Obviously, terrestrial light will match the Earth's movement through space but not the source's movement on Earth. This means that Earth's mass or gravitational field quickly changes the speed of light to match its own movement.

As light leaves a star, it will travel at some speed close to *c* relative to that star's motion. The light must continue to travel at that speed while it is in that star's gravitational influence. As the light leaves the proximity of its source, the gravity from that source will weaken allowing the light to propagate faster. Gravity from other sources will start to influence the light wave. That electromagnetic wave will start to speed up or slow down to match the combination of gravitational forces from each environment that it passes through. This means that the two light beams from our binary star example will eventually change speed so that they are both propagating together, at the same speed, as they travel through the varying gravitational fields.

As that same light wave approaches our Sun's gravitational field, it must change its speed to match the Sun's path through space. By the time an observer measures this light on Earth, the light apparently will match the Earth's gravitational field. Because of this, all light speed measurements on Earth will look the same from any direction.

Once this idea of the propagation of light matching the local gravitational field is given some merit, a whole new set of logical questions will come to mind: How far from Earth does light start to match Earth's movement through space? Satellites rotate around Earth. Do the satellites have enough of a gravitational influence of their own to cause light to try to match their speed, or does the Earth's gravitational field overwhelm it? How far from Earth does light have to travel before it changes its speed to match the Sun's gravitational field? How far from the Sun does light have to travel before the next star's gravitational field or our galaxy's combined gravitational field takes over? Does the gravitational field from multiple objects simply average out to control the speed of light in that area of space?

What happens to light as it leaves the strong gravitational influences of galaxies? We have seen that light slows down near a strong gravitational field causing the light waves to bend. Our own Sun demonstrates that light leaves its source with a speed that is relative to the source's movement through space and therefore relative to its gravitational field. We have seen that light changes its speed to match the observer's gravitational field. What can we infer is happening to the speed of light when the gravitational field gets extremely weak? Considering the gravitational lensing effect of a large cluster of galaxies, it is clear that light is continuing to change its speed at long distances (at least one million-light-years) from the cluster. This strongly suggests that light may continue to speed up, the weaker the gravitational field gets. Over enough time, as light travels between galaxies, a weak field might even degrade light in some way.

It is possible that light degrades over time as it travels between galaxies. Degrading (or tired) light could explain why the universe appears to be expanding. If the universe actually is expanding like current theory predicts, we would be doomed to a future of darkness in the sky. The further we look back in time, the younger the objects are in the universe. The universe is obviously changing and getting older, there is no doubt about that. I would have liked to think that eventually everything would again collapse upon itself, as all galaxies sink into their own black holes. All the black holes would hopefully collapse inward to some super massive object. When that massive universe-full object reached its own critical mass, maybe it would trigger another Big Bang and start everything over again; a cyclical universe that expands and contracts. If this is not going to happen, then where did our first Big Bang come from? We can only wonder.

The mistakes made by Einstein and the other mathematicians leave room for much speculation about reality—questions left open for future generations to potentially solve with novel techniques. Until then, try not to let your own frame of reference cloud your ability to explore reality in an unbiased and logical manner.

About the Author:

John Paul Keyes was born in California in 1951 and grew up in Oregon. After completing his BS degree in Mathematics at the University of Washington he founded a software company that develops and sells business applications all across the world.

His greatest wish is that the world sees where Einstein went wrong so that science can progress forward with a logical foundation.

Made in the USA
Monee, IL
04 November 2026

67377425R00036